Mermaids on a Mission to Save the Oceans

Written by: Janet Balletta

Illustrations by: Alyssa Figueroa & Alexander C. Appello

Mermaids on a Mission to Save the Oceans
©2018 by Janet Balletta

Book edited by Ana Morris
Front cover by Alexander C. Appello
Title page & Back cover by Alyssa Figueroa
Interior artwork collaboration of Alyssa Figueroa & Alexander C. Appello

WRB Publishing
Palm City, FL 34990
wrb1174@gmail.com

ISBN-13: 978-0-9909040-7-6

This book is dedicated to children who can make

positive changes to save the ocean habitats of our

blue planet. May you enjoy these pearls of wisdom.

Marlene delighted in watching the sunset on Miami Beach. She looked forward to it no matter how hard her day had been. She loved the way the vibrant colors lit the sky and reflected on the water. It was the perfect way to end a day of protecting the Atlantic Ocean. Marlene was one of the mermaids chosen by the Council for the Mermaid Mission to save the oceans from mankind's pollution. One mermaid was selected for each of the seven continents. Marlene was assigned to North America.

Marlene was careful to swim beyond the buoys so humans on the shore could not see her. She flexed her sparkling tail as she glided in the water waiting for the celestial show to begin. It was her favorite time of day because she could unwind in a sea of warm bubbly foam. She closed her eyes and breathed in the salty aroma. Marlene smiled as she thought about the upcoming reunion with her six mermaid sisters from around the globe. She couldn't wait to see her sisters as they always had a fabulous time together.

Once home, Marlene remembered to call her sister, Fina. She was in South America guarding the oceans in that part of the world. They spoke almost daily sharing fresh ideas on keeping the oceans clean. Marlene was grateful Fina had helped her stop the chemical dumping in the Gulf of Mexico. That big mess hurt the rivers, oceans, and animals in the area. Marlene reached for her pink clam phone and blurted, "Call Fina."

Fina was home dancing to salsa music swaying her glistening red hair from side to side. She answered her clam phone before it rang using her superpowers.

"Hi, Fina, guess what?" asked Marlene. "We're getting medals from the Council for stopping the chemical dumping in the Gulf of Mexico," Marlene said.

"Wow, that's marvelous!" cried Fina. "Have you told the others yet?" Fina asked.

"No, we'll tell them at our reunion in Asia next week," Marlene giggled.

Marlene updated Fina on how she was persuading humans to stop putting oil, plastics, and chemicals into the ocean; and to conserve water and increase recycling in Florida. She had recruited kids to remind their families and friends that those things polluted lakes and rivers, destroyed animal habitats, and harmed the planet. At first, it was hard getting humans to understand that all living things depended on one another to survive. Now, they were mindful of the need for clean water to live on the blue planet. The kids were doing a great job of spreading the message.

Fina opened Aqua Pura, a water park in South America that was really a secret hospital for animals. Many animals were taken there to be nursed back to health. Sea turtles went there to get plastic soda wrappers removed from their necks. Birds and mammals went there to get cleaned up after an oil spill. Oil destroyed their feathers and fur that kept them from freezing. Fina worked hard to take care of her sea friends. She was the first to receive a medal from the Council for opening the animal hospital in South America.

In Africa, Jazzy guarded the oceans waters while humans drilled for oil. Her shark friends aided her by leading the animals out of harm's way and then escorting them back home when the drilling was finished. The sharks loved patrolling the seas with Jazzy. Her dark shiny hair flowed freely through the waves leading the way to safety. In other parts of Africa, where water was scarce, Jazzy helped them install water wells to increase the water supply on that continent. This made life much easier for humans in Africa.

In Asia, Marbella protected the seas by convincing humans in city hall to change the laws to reduce water pollution. She had a gift for getting humans to do anything she asked with her mesmerizing smile. Her job was vital because of the large amounts of seafood imported to the United States from Asia. Marbella became furious whenever humans got sick from eating spoiled fish. After a long day, she came home and relaxed in her spa to forget about work for a while.

Belinda oversaw the oceans in Australia. She adored animals, nature, and being outdoors. Her glittering scales were like stars twinkling in the deep blue sea. Belinda had a talent for making her own medicines using plants like seaweed, kelp, and coral. When things were running smoothly in her part of the world, she swam to Aqua Pura and brought medicine to the animal hospital. Belinda took great pride in making her medicines and teaching humans the ocean was a natural resource with enough plants to heal the whole planet.

Milana was assigned to Europe. She managed an International School of Languages where she taught humans how to keep the ocean waters clean in any language. Milana had a knack for languages and could speak mermaid Spanish, Italian, French, and Greek. She had the sweetest voice that echoed through the water like musical notes on a harp. Whenever she wasn't busy teaching or watching over the Mediterranean, she took care of the red crabs that seemed to multiply overnight.

Ava secured the frozen ocean in Antarctica which worked well, because she was fond of the cool weather and adored the penguins. She had a flair for communicating with the arctic animals and didn't like leaving them for long. Ava was always the first one to return home after a reunion. Her job was very demanding, with so many glaciers melting and drifting into the ocean. The Emperor Penguins helped her keep track of the sliding glaciers, which was a full-time job now with global warming.

The Council was a group of twelve elder mermaids and mermen who broadcast the global news around the clock. They gave the seven mermaids superpowers and the ability to change into humans to fulfill their mission. Each mermaid's job was tuning into the news day and night using their superpowers. They listened for anything that might harm the oceans surrounding their continents. Many wished to be part of the Council but only the finest were selected for the Mermaid Mission.

Meanwhile back in North America, Marlene had finished talking to Fina and was putting the clam phone down when she felt the ocean floor tremble. Suddenly, a strong rip current spiraled around Marlene causing her to lose her balance for a moment. Then using her superpowers, she broke free and tuned into the news. Marlene glimpsed her teacups and saucers floating in the water from the corner of her eye while listening to the news. The news report stated they were expecting more tremors due to the oil drilling.

Marlene swished outside of her apartment to check on the animals and make sure they were safe. Her dolphin friends were already on the job soothing the scared fish who became nervous at the slightest commotion in the ocean. The fish waited for the dolphins to send the all-clear vibration that rippled through the water for hundreds of miles. Once the fish got the all-clear, they came out of their hiding places.

Marlene dashed around the Atlantic Ocean like speed lightning using her superpowers. She had to make sure that everything was fine. Then she swam home to rest for a while, as she had a feeling it was going to be a long night. At home, she made herself a cup of green tea and fell asleep on her pink clam lounger. Luckily, there were no more tremors that night.

In the morning, the news reported calmer waters for the next few days which was perfect for her trip to go see her mermaid sisters in Asia. That afternoon, Marlene finished preparing for her trip and met with the dolphins. She always informed them when she was going away. They were thrilled to help protect the Atlantic Ocean whenever she left the continent. Marlene never returned home without their favorite dolphin treats to enjoy.

In Asia, Marlene shared the surprise with her mermaid sisters. Soon the Council arrived and Charles, the eldest member, started the meeting.

"We've come to present you with a Medal of Excellence for the fine job you are doing of protecting the oceans from mankind's pollution!" he exclaimed. Then, he placed a medal on each mermaid.

Alyssa and Gabriela, the youngest Council members, sang the merfolk anthem and the party commenced. There were scrumptious foods, merry dancing, and loud merfolk laughter. Best of all, the ocean animals were happy and there was harmony in the deep-blue sea.

Teacher's Guide

Vocabulary:

Buoy	(noun)	A floating object anchored in a body of water to warn of danger.
Broadcast	(adjective)	The act of sending sounds or images by radio or television.
Celestial	(adjective)	Something related to the sky or the heavens.
Chemical	(noun)	A substance that is harmful to the environment if misused like pesticides on food.
Glistening	(verb)	To shine, glow, or sparkle.
Habitat	(noun)	A place where animals live.
Mesmerizing	(adjective)	To hypnotize or spellbind.
Scarce	(adjective)	Something lacking in quantity or not enough.

Questions:

1. Why is the Mermaid Mission important?

2. How does chemical dumping hurt the lakes, rivers, and oceans?

3. What happened after Marlene felt the ocean floor tremble?

4. Why does Jazzy install water wells in Africa?

5. What does Belinda do with the seaweed, kelp and coral?

6. How does plastic harm the oceans?

7. What happened before Marlene left for her trip to meet her sisters?

8. How would you persuade humans to care for the oceans and blue planet?

9. What are the names of the seven continents and five oceans?

10. What science stem project could you come up with to solve mankind's pollution?

11. Visit my website Janetballetta.com for cool science experiments to go with the book.

www.ingramcontent.com/pod-product-compliance
Lightning Source LLC
Chambersburg PA
CBHW061152030426
42336CB00002B/26